PRACTICE EXAMS
FOR JUNIOR
MATH OLYMPIADS

BOOK 1

PRACTICE EXAMS FOR JUNIOR MATH OLYMPIADS

BOOK 1

Roman Kvasov, Ph.D.

Published by 42 Points. San Juan, Puerto Rico, 2024.

CONTENTS

Dedicated to all students who love Math Olympiads

INTRODUCTION

The journey to becoming a successful Junior Math Olympiad participant is one of dedication, practice, and a love for mathematics. This book, "Practice Exams for Junior Math Olympiads," is designed to provide additional practice exams for students training for various Junior Math Olympiads such as the United States of America Junior Math Olympiad (USAJMO), Canadian Junior Mathematical Olympiad (CJMO), Korean Junior Math Olympiad (KJMO), Junior Balkan Math Olympiad (JBMO), and the Central America and the Caribbean Math Olympiad (OMCC).

The problems in this book cover fundamental concepts in algebra, geometry, number theory, and combinatorics, carefully selected with students in grades 8-10 in mind. These students may already have some exposure to Math Olympiad topics and are looking to deepen their understanding and problem-solving skills.

Math Olympiad problems are typically referred to by their number, such as P1 for Problem 1, P2 for Problem 2, and so on. Each of the six exams in this book includes detailed solutions to help students understand the reasoning and methods behind each problem.

The difficulty of the problems is structured as follows: P1 is the easiest and P6 is the hardest problem on each test. P4 is slightly harder than P1 but easier than P2. P2 is easier than P5, which in turn is easier than P3:

$$P1 < P4 < P2 < P5 < P3 < P6$$

There are two recommended ways to work through the problems in this book. One method is to use the exams as mock exams and complete them in one sitting, just like a real Math Olympiad test. This approach helps build stamina and simulates the actual competition experience. The other method is to gradually work through the problems, starting with P1 and P4, then moving to P2 and P5, and finally tackling P3 and P6. This gradual approach allows students to build their skills progressively.

Although some Junior Math Olympiads, such as the Canadian Junior Mathematical Olympiad (CJMO) or Junior Balkan Math Olympiad (JBMO), have different formats and numbers of problems, this book still provides practice material that reflects the level of these competitions. By working through these problems, students will be well-prepared for the challenges they will face in their respective Olympiads.

The author wishes the reader the best of luck on their Math Olympiad journey and hopes that they will find this book both challenging and enjoyable. Embrace the process, learn from each problem, and most importantly, have fun with mathematics.

Roman Kvasov, Ph.D.

CHAPTER 1

PRACTICE EXAM #1

Problems for Day 1

Problem 1

Is it possible that the number of the form $8^n + 1$ is divisible by the number of the form $8^k - 1$, for some positive integers n and k?

Problem 2

Prove the inequality for positive values of the variables

$$\frac{a}{2b + 7c} + \frac{b}{2c + 7a} + \frac{c}{2a + 7b} \geq \frac{1}{3}$$

Problem 3

Circle ω_1 is tangent to the sides AB and BC of the triangle ABC at the points D and E, respectively, and is tangent to its circumcircle internally at the point F. Prove that the line DE passes through the incenter I of the triangle ABC.

Practice Exams For Junior Math Olympiads - Book 1
by Roman Kvasov, Ph.D.

11

Problems for Day 2

Problem 4

Points D, E and F are the midpoints of the sides AB, BC and CA of a right triangle ABC with $\angle ACB = 90°$. Equilateral triangles AFX and BEY are constructed on the segments AF and BE, such that they lie completely outside of the triangle ABC. Find the angle $\angle DXY$.

Problem 5

Solve the equation in real numbers

$$2025^x + 4^x + 1 = 90^x + 45^x + 2^x$$

Problem 6

Initially the numbers 14 and 15 are written on the board. It is allowed to take any written number n, erase it, and write three positive integers a, b and c, such that

$$abc = 3n^3$$

Prove that if at some moment there are exactly 2000 numbers written on the board, then one of them is not greater than 150.

Solutions for Day 1

Solution for Problem 1

Answer: no.

Let us assume that there exists positive integers n and k, such that $8^n + 1$ is divisible by $8^k - 1$.

We will work modulo 7.

Start by noticing that

$$8^k - 1 \equiv (1)^k - 1 \equiv 1 - 1 \equiv 0 \pmod{7}$$

and consequently, $8^k - 1$ is divisible by 7.

However, since

$$8^n + 1 \equiv (1)^n + 1 \equiv 1 + 1 \equiv 2 \pmod{7}$$

then, $8^n + 1$ is not divisible by 7. This implies that $8^n + 1$ cannot be divisible by $8^k - 1$, and we obtained a contradiction.

We conclude that no number of the form $8^n + 1$ is divisible by the number of the form $8^k - 1$, as desired.

Solution for Problem 2

Let us start by proving the following Lemma.

Lemma

For positive a, b and c

$$a^2 + b^2 + c^2 \geq ab + bc + ac$$

Proof

Let us apply the AM-GM Inequality to the following expression

$$2\left(a^2 + b^2 + c^2\right) = 2a^2 + 2b^2 + 2c^2$$
$$= \left(a^2 + b^2\right) + \left(b^2 + c^2\right) + \left(a^2 + c^2\right)$$
$$\geq 2ab + 2bc + 2ac$$
$$= 2\left(ab + bc + ac\right)$$

which implies that

$$a^2 + b^2 + c^2 \geq ab + bc + ac$$

and the lemma is proven.   ∎

Let us rewrite the right-hand side of the needed inequality

$$\frac{a}{2b + 7c} + \frac{b}{2c + 7a} + \frac{c}{2a + 7b}$$

as follows:

$$\frac{a^2}{2ab + 7ac} + \frac{b^2}{2bc + 7ab} + \frac{c^2}{2ac + 7bc}$$

Now we will apply the Titu's Lemma to the following tuples of numbers: (a, b, c) and $(2ab + 7ac, 2bc + 7ab, 2ac + 7bc)$:

$$\frac{a^2}{2ab + 7ac} + \frac{b^2}{2bc + 7ab} + \frac{c^2}{2ac + 7bc} \geq \frac{(a + b + c)^2}{9(ab + bc + ac)}$$

Notice that it will be enough to prove that

$$\frac{(a + b + c)^2}{9(ab + bc + ac)} \geq \frac{1}{3}$$

This inequality is equivalent to the following:

$$3\left(a + b + c\right)^2 \geq 9\left(ab + bc + ac\right)$$
$$\left(a + b + c\right)^2 \geq 3\left(ab + bc + ac\right)$$
$$a^2 + b^2 + c^2 + 2ab + 2bc + 2ac \geq 3ab + 3bc + 3ac$$
$$a^2 + b^2 + c^2 \geq ab + bc + ac$$

which holds by the lemma, as desired.

Solution for Problem 3

The solution presented below refers to Figure 1.1.

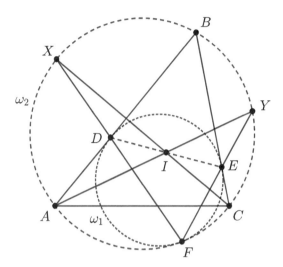

Figure 1.1 Application of the Pascal's Theorem to the hexagon $ABCXFY$ in Problem 3.

Let ω_2 be the circumcircle of the triangle ABC, and let X and Y be the midpoints of the arcs AB and BC, respectively. Notice that AI passes through Y, and CI passes through X.

Let us prove that FE passes through Y, and FD passes through X.

Consider the homothety with center F that maps the circle ω_1 to the circle ω_2. Then, it maps the line AB to the line parallel to AB and tangent to ω_2. Therefore, it maps D to the midpoint of the arc AB, i.e. to the point X. From here FD passes through X.

Similarly, this homothety maps the line BC to the line parallel to BC and tangent to ω_2. Therefore, it maps E to the midpoint of the arc BC, i.e. to the point Y. From here FE passes through Y.

Notice that the points I, D and E are the intersections of the lines AY and CX, AB and FX, BC and FY, respectively. From here by the Pascal's Theorem applied to the nonconvex hexagon $ABCXFY$, it follows that the points I, D and E are collinear, and consequently, DE passes through I, as desired.

Solutions for Day 2

Solution for Problem 4

The solution presented below refers to Figure 1.2.

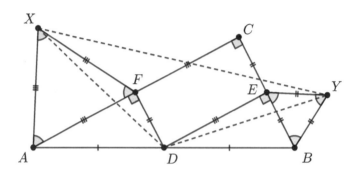

Figure 1.2   Triangles XFD and YED are congruent in Problem 4.

Answer: $\angle DXY = 30°$.

First, let us prove that $DX = DY$.

Start by noticing that DE and DF are the midsegments of the triangle ABC. This implies that

$$XF = FA = \frac{AC}{2} = DE$$

and also

$$YE = EB = \frac{BC}{2} = DF$$

Since the triangles AFX and BEY are equilateral, and $DE \perp BC$ and $DF \perp AC$, then

$$\angle XFD = \angle XFA + \angle AFD = 90° + 60° = 150°$$

and also

$$\angle YED = \angle YEB + \angle BED = 90° + 60° = 150°$$

From here the triangles XFD and YED are congruent by the Side-Angle-Side Postulate. Therefore $\angle FXD = \angle EDY$, $\angle XDF = \angle EYD$ and $DX = DY$.

Notice that this implies that the triangle XDY is isosceles, and consequently, $\angle DXY = \angle DYX$.

Now let us show that $\angle XDY = 120°$.

Indeed, we have

$$\begin{aligned}
\angle XDY &= \angle XDF + \angle EDF + \angle EDY \\
&= \angle XDF + 90° + \angle FXD \\
&= (\angle XDF + \angle FXD) + 90° \\
&= (180° - \angle XFD) + 90° \\
&= (180° - 150°) + 90° \\
&= 120°
\end{aligned}$$

Therefore, from the triangle XDY, we have $\angle DXY = 30°$, as desired.

Solution for Problem 5

Answer: $x = 0$.

Let us start b rewriting the given equation as follows:

$$2025^x + 4^x + 1 - 90^x - 45^x - 2^x = 0$$

and making the substitutions $a = 45^x$ and $b = 2^x$.

We will now rewrite this equation in new variables, multiply it by 2 and complete the squares:

$$\begin{aligned}
2025^x + 4^x + 1 - 90^x - 45^x - 2^x &= 0 \\
a^2 + b^2 + 1 - ab - a - b &= 0 \\
2a^2 + 2b^2 + 2 - 2ab - 2a - 2b &= 0 \\
\left(a^2 - 2a + 1\right) + \left(b^2 - 2b + 1\right) + \left(a^2 - 2ab + b^2\right) &= 0 \\
(a - 1)^2 + (b - 1)^2 + (a - b)^2 &= 0
\end{aligned}$$

From here we have that

$$\begin{aligned}
a - 1 &= 0 \\
b - 1 &= 0 \\
a - b &= 0
\end{aligned}$$

This implies that $a = 1$ and $b = 1$. Therefore, we have $45^x = 1$ and $2^x = 1$. The last two equations hold only for $x = 0$, as desired.

Solution for Problem 6

Let $x_1, x_2, \ldots, x_m$ be the numbers on the board, and let us consider the sums of cubes of their reciprocals:

$$S = \sum_{i=1}^{m} \left(\frac{1}{x_i^3} \right)$$

Notice that since $abc = 3n^3$, then by the AM-GM Inequality

$$\frac{1}{a^3} + \frac{1}{b^3} + \frac{1}{c^3} \geq \frac{3}{abc} = \frac{3}{3n^3} = \frac{1}{n^3}$$

From here we see that if the number n was erased, then the quantity S is not decreasing, and consequently, is a monovariant.

Let z be the smallest number on the board when there are exactly 2000 numbers written, i.e. $z \leq x_i$ for all $i = 1, 2, \ldots, 2000$. Then, we have

$$\frac{2000}{z^3} \geq \sum_{i=1}^{2000} \left(\frac{1}{x_i^3} \right) \geq \frac{1}{14^3} + \frac{1}{15^3} \geq \frac{2}{15^3}$$

This implies that

$$\frac{2000}{z^3} \geq \frac{2}{15^3}$$

and consequently

$$z \leq 150$$

as desired.

CHAPTER 2

PRACTICE EXAM #2

Problems for Day 1

Problem 1

Solve the equation in positive integer numbers

$$5^m - 2^n = 3$$

Problem 2

Consider a white 4×4 grid consisting of 16 unit squares. How many different ways are there to color one or more unit squares gray, such that the resulting gray area forms a rectangle?

Problem 3

Find all functions $f : \mathbb{Z} \to \mathbb{R}$ that satisfy the equation for all integer x and y:

$$f(x+y) + f(xy-1) = (f(x)+1)(f(y)+1)$$

Practice Exams For Junior Math Olympiads - Book 1
by Roman Kvasov, Ph.D.

19

Problems for Day 2

Problem 4

Let D be a point on the median BM of triangle ABC. Through point D, a line is drawn parallel to side AB, and through point C, a line is drawn parallel to the median BM. The two drawn lines intersect at point E. Prove that $BE = AD$.

Problem 5

Positive integers m and n satisfy the equation

$$\frac{m+n}{2} = \sqrt{mn} + 1$$

Determine all pairs (m, n) of such numbers.

Problem 6

Teacher has a very long strip of paper and a 2019-digit number n. He writes out consecutive natural numbers on a strip in a row without spaces, starting with n:

$$n, n+1, n+2, \ldots$$

Prove that at some point the number on the strip will be divisible by 101.

Solutions for Day 1

Solution for Problem 1

Answer: $m = n = 1$.

Let us rewrite the given equation as follows:

$$2^n = 5^m - 3$$

We will proceed by doing the following casework.

- If $n = 1$, then the equation becomes

$$5 = 5^m$$

  which implies that $m = 1$.

- If $n \geq 2$, then the left-hand side is divisible by 4, and so should be the right-hand side. However

$$5^m - 3 \equiv (1)^m - 3 \equiv (1) - 3 \equiv -2 \pmod 4$$

  and we obtained a contradiction.

We conclude that the only positive integer solutions of the equation are $m = n = 1$, as desired.

Solution for Problem 2

Answer: total number of possible gray rectangles is 100.

Start by noticing that selecting the bottom-left and top-right squares uniquely determines a rectangle (see Figure 2.1).

For all $1 \leq i, j \leq 4$, let us assign the coordinates (i, j) to the unit squares, such that $(1, 1)$ and $(4, 4)$ are the bottom-left and the top-right squares, respectively.

Notice that for each choice of the top-right square (i, j), there are exactly ij possible choices for the bottom-left square. Therefore, the total number of gray rectangles equal to the sum of all possible products of two numbers from the set $\{1, 2, 3, 4\}$:

$$\sum_{1 \leq i, j \leq 4} ij$$

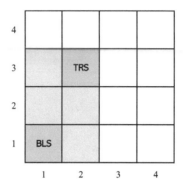

Figure 2.1   Bottom-left square (**BLS**) and top-right square (**TRS**) uniquely determine a rectangle in Problem 2.

This sum is equal to

$$(1 + 2 + 3 + 4) \cdot (1 + 2 + 3 + 4) = 10 \cdot 10 = 100$$

as desired.

Solution for Problem 3

First, substituting successively $x = 0$, $y = 0$ and $x = 0$, $y = -1$, we obtain the following system of equations:

$$f(0) + f(-1) = (f(0) + 1)^2$$
$$2f(-1) = (f(0) + 1)(f(-1) + 1)$$

from where we find that $f(0) = 0$ and $f(-1) = 1$.

Now substituting $x = -1$, $y = -1$ we have

$$f(-2) + f(0) = (f(-1) + 1)^2$$

from where we find that $f(-2) = 4$.

And finally, substituting $x = 1$, $y = -1$, we obtain

$$f(0) + f(-2) = (f(1) + 1)^2$$

from where we find that $f(1) = 1$.

Now let us substitute $x = n$, $y = 1$:

$$f(n+1) + f(n-1) = ((f(n)+1)\,((f(1)+1)$$
$$f(n+1) + f(n-1) = 2\,((f(n)+1)$$
$$f(n+1) = 2(f(n)+1) - f(n-1) \tag{2.1}$$

For the nonnegative integers n, we will use induction to prove that $f(n) = n^2$.

The basis of induction holds, since $f(0) = 0$ and $f(1) = 1$.

For the inductive step, let us assume that $f(k) = k^2$ for some integer k, and prove that $f(k+1) = (k+1)^2$. Substituting $n = k$ into the equality (2.1) we have

$$\begin{aligned}
f(k+1) &= 2(f(k)+1) - f(k-1) \\
&= 2(k^2+1) - (k-1)^2 \\
&= 2k^2 + 2 - k^2 + 2k - 1 \\
&= k^2 + 2k + 1 \\
&= (k+1)^2
\end{aligned}$$

For the nonpositive integers n, we will rewrite the equality (2.1) in the form

$$f(n-1) = 2\,(f(n)+1) - f(n+1) \tag{2.2}$$

Let us put $n = -m$ where m is a nonnegative integer and substitute it into the equality (2.2):

$$f\,(-(m+1)) = 2\,(f(-m)+1) - f(-(m-1)) \tag{2.3}$$

We will use induction to prove that $f(-m) = m^2$ for all nonnegative integers m.

The basis of induction holds, since $f(0) = 0$ and $f(-1) = 1$.

For the inductive step, let us assume that $f(-k) = k^2$ for some integer k, and prove that $f\,(-(k+1)) = (k+1)^2$. Substituting $m = k$ into the equality (2.3) we have

$$\begin{aligned}
f\,(-(k+1)) &= 2\,(f(-k)+1) - f(-(k-1)) \\
&= 2(k^2+1) - (k-1)^2 \\
&= 2k^2 + 2 - k^2 + 2k - 1 \\
&= k^2 + 2k + 1 \\
&= (k+1)^2
\end{aligned}$$

We conclude that $f(x) = x^2$ for all $x \in \mathbb{Z}$, as desired.

Solutions for Day 2

Solution for Problem 4

The solution presented below refers to Figure 2.2.

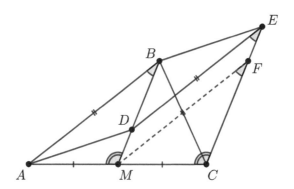

Figure 2.2   Point F is the intersection of the line CE and the line through the point M parallel to DE in Problem 4.

Let us draw the line through the point M parallel to the line DE and the point F being its intersection with the line CE. Notice that from here we have $MF = DE$.

Since $AB \parallel DE \parallel MF$ and $BM \parallel CF$, then

$$\angle AMB = \angle MCF$$

and

$$\angle ABM = \angle DEF = \angle MFC$$

From here the triangles AMB and MCF are congruent by the Angle-Angle-Side Postulate, and therefore $AB = MF$. Taking into account that $MF = DE$, we have $AB = DE$. This implies that the quadrilateral $ABED$ is a parallelogram, and $BE = AD$, as desired.

Solution for Problem 5

Answer: $m = 2k^2$, $n = 2(k-1)^2$ or $m = 2(k-1)^2$, $n = 2k^2$, where $k \in \mathbb{N}$.

Without loss of generality, we can assume that $m \leq n$.

We will make a substitution $n = m + d$, where d is some nonnegative integer. The given equality becomes

$$\frac{m + n}{2} - \sqrt{mn} = 1$$

$$\frac{2m + d}{2} - \sqrt{m(m + d)} = 1$$

$$\frac{2m + d}{2} - 1 = \sqrt{m(m + d)}$$

$$\frac{2m + d - 2}{2} = \sqrt{m(m + d)}$$

$$2m + d - 2 = 2\sqrt{m(m + d)}$$

Let us now square both sides and rearrange the terms as follows:

$$(2m + d - 2)^2 = \left(2\sqrt{m(m + d)}\right)^2$$

$$4m^2 + 4md - 8m + d^2 - 4d + 4 = 4m^2 + 4md$$

$$d^2 - 4d - 8m + 4 = 0$$

$$(d - 2)^2 = 8m$$

This implies that d is even, i.e., $d = 2n$ for some $n \in \mathbb{N}$. Then the equation becomes

$$2m = (n - 1)^2$$

which implies that n is odd, i.e., $n = 2k - 1$ for some $k \in \mathbb{N}$.

From here

$$m = 2k^2$$

$$n = 2(k - 1)^2$$

satisfy the original equation for all $k \in \mathbb{N}$, as desired.

Solution for Problem 6

Let x be the number on the strip after the last 2021-digit number is written. Notice that the next number to be written will be the number $y = 10^{2021}$ which will contain exactly 2022 digits.

Now we will write the numbers in groups of two consecutive numbers. Let us count the contribution of each group and the number x modulo 101.

Let us write a numbers, where $a = 2k$ and k is the number of groups. Then the number written on the board will be equal to

$$10^{2022a} \cdot x + 10^{2022(a-1)} \cdot (y) + 10^{2022(a-2)} \cdot (y+1) + \dots$$

$$\dots + 10^{2022} \cdot (y+a-2) + (y+a-1)$$

Since

$$10^2 = 100 \equiv -1 \pmod{101}$$

then

$$10^{2022a} \cdot x \equiv 10^{2022 \cdot 2k} \cdot x$$

$$\equiv \left(10^2\right)^{2022k} \cdot x$$

$$\equiv (-1)^{2022k} \cdot x$$

$$\equiv x \pmod{101}$$

This implies that the contribution of x into the number on the strip is invariant modulo 101.

Let us now count the contribution of each group. Notice that

$$10^{2022} = 100^{1011} \equiv (-1)^{1011} \equiv -1 \pmod{101}$$

Therefore, the consecutive powers of 10^{2022} are congruent to $-1, +1, -1, +1$, etc. This implies that after each group of two numbers is written the remainder of number increases by 1 modulo 101. Indeed, the contribution of each group is congruent to

$$(-1) \cdot (y+i-1) + (1) \cdot (y+i) \equiv -y - i + 1 + y + i \equiv 1 \pmod{101}$$

Since the remainder increases by 1, then it will eventually become a multiple of 101, as desired.

CHAPTER 3

PRACTICE EXAM #3

Problems for Day 1

Problem 1

In an acute triangle ABC the measure of the angle B is equal $60°$. The altitudes CE and AD intersect at H. Show that the circumcenter O of the triangle ABC lies on the common angle bisector of the angles $\angle AHE$ and $\angle CHD$.

Problem 2

A pair of positive integer numbers (a, b), is called *awesome* if $a > b$ and the least common multiple of the numbers a and b is divisible by $a - b$. It is known that for some number n, among all its positive integer divisors, there is only one *awesome* pair. Find n.

Problem 3

Given positive numbers a, b, c, such that $abc = 1$ and

$$a + b + c > \frac{1}{a} + \frac{1}{b} + \frac{1}{c}$$

Prove that exactly one of the numbers a, b, c is greater than 1.

Practice Exams For Junior Math Olympiads - Book 1
by Roman Kvasov, Ph.D.

27

Problems for Day 2

Problem 4

Show that the primes of the form

$$2^{2^s} + 1$$

where $s \in \mathbb{N}$, cannot be represented as a difference of cubes of two positive integers.

Problem 5

Given a prime $p > 2$. N boys are sitting around a circle table. Each boy has a positive integer that he divides by p, and then writes the remainder on a piece of paper. Then each boy writes on his piece of paper the remainder of the division by p of the product of his number and the number of his right neighbor. Find the maximum value of N if it is known that the first numbers written on the papers are all different and every piece of paper has two different numbers.

Problem 6

Prove that for any natural number $n \geq 2$ and for any real numbers $a_1, a_2, \ldots, a_n$ satisfying

$$a_1 + a_2 + \ldots + a_n \neq 0$$

the following equation has at least one real root:

$$\sum_{i=1}^{n} a_i \prod_{j \neq i} (x - a_j) = 0$$

Solutions for Day 1

Solution for Problem 1

The solution presented below refers to Figure 3.1.

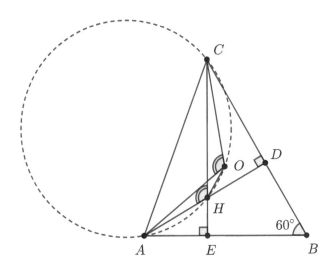

Figure 3.1 Quadrilateral $AHOC$ is cyclic in Problem 1.

Without loss of generality let us assume that the point O lies in the triangle CHD.

From the quadrilateral $BEHD$ we have that $\angle EHD = 120°$. From here we have that $\angle AHC = 120°$ and $\angle CHD = 60°$.

Since $\angle AOC$ is central, then it equals twice the angle $\angle ABC$, and therefore $\angle AOC = 120°$.

Since $\angle AOC = \angle AHC$, then the quadrilateral $AHOC$ is cyclic. Since the triangle AOC is isosceles and $\angle AOC = 120°$, then $\angle ACO = 30°$.

Since $AHOC$ is cyclic, then $\angle DHO = \angle ACO = 30°$ and OH is the angle bisector of the angle $\angle CHD$, as desired.

Solution for Problem 2

Answer: $n = 2$.

It is not hard to check that $n = 2$ works, while $n = 1$ does not work.

For $n \geq 3$ we will proceed by doing the following casework.

- Let n be an even number $n = 2k$, where $k \geq 2$, $k \in \mathbb{N}$. Then n has the following divisors: 1, 2, k and $2k$. The pairs $(1, 2)$ and $(k, 2k)$ are distinct *awesome* pairs, and therefore, no even n works.

- Let n be an odd number greater than 1 and (a, b) be an *awesome* pair of its divisors. Then $\mathrm{lcm}\,(a, b)$ is odd and cannot be divisible by $a - b$, which is even.

We conclude that the only such positive integer n is $n = 2$, as desired.

Solution for Problem 3

Taking into account that $abc = 1$, let us start by rewriting the given inequality as follows:

$$a + b + c > \frac{1}{a} + \frac{1}{b} + \frac{1}{c}$$

$$a + b + c > \frac{ab + bc + ac}{abc}$$

$$a + b + c > ab + bc + ac \tag{3.1}$$

Let us now consider the polynomial $P(x)$ defined as

$$P(x) = (x - a)(x - b)(x - c)$$

Notice that $P(x)$ is of degree 3 and has three positive roots at $x = a$, $x = b$ and $x = c$.

From the Vieta's Formulas we have

$$P(x) = x^3 - (a + b + c)x^2 + (ab + bc + ac)x - abc$$

and taking into account that $abc = 1$

$$P(x) = x^3 - (a + b + c)x^2 + (ab + bc + ac)x - 1$$

Let us consider the value of $P(1)$:

$$P(1) = 1 - (a + b + c) + (ab + bc + ac) - 1 = (ab + bc + ac) - (a + b + c)$$

From the equality (3.1) we have

$$P(1) = (ab + bc + ac) - (a + b + c) < 0$$

Notice that the inequality $P(1) < 0$ is equivalent to

$$(1 - a)(1 - b)(1 - c) < 0$$

The last inequality implies that either all three or exactly one of the numbers a, b, c is greater than 1. If all numbers a, b and c are greater than 1, then $abc > 1$, which contradicts the condition $abc = 1$.

We conclude that exactly one of the numbers a, b, c is greater than 1, as desired.

Solutions for Day 2

Solution for Problem 4

Let us assume that the exists $s \in \mathbb{N}$, such that $2^{2^s} + 1$ is prime and can be represented as a difference of cubes of two positive integers, i.e.

$$2^{2^s} + 1 = a^3 - b^3$$

for some $a, b \in \mathbb{N}$.

Let us factor the right-hand side of the last equality as follows

$$2^{2^s} + 1 = (a - b)\left(a^2 + ab + b^2\right)$$

Notice that since

$$a^2 + ab + b^2 \geq 1 + 1 + 1 = 3$$

and $2^{2^s} + 1$ is prime, then $a - b = 1$. From here we have $a = b + 1$, and we can substitute it into the initial equation:

$$2^{2^s} + 1 = a^3 - b^3$$
$$2^{2^s} + 1 = (b + 1)^3 - b^3$$
$$2^{2^s} + 1 = b^3 + 3b^2 + 3b + 1 - b^3$$
$$2^{2^s} = 3b^2 + 3b$$
$$2^{2^s} = 3(b^2 + b)$$

Since the right-hand side of the last equation is divisible by 3, but the left-hand side is not, then we obtained a contradiction.

We conclude that no prime of the form $2^{2^s}+1$ can be represented as a difference of cubes of two positive integers, as desired.

Solution for Problem 5

Answer: the maximum value of N is $p-2$.

Notice that none of the first numbers is 0, since then the boy who has 0 will have to write it twice on his piece of paper.

Also notice that none of the first number is 1, since then the left neighbor of the boy who has 1 will have to write his number twice on his piece of paper.

From here we have that $N \leq p-2$. Let us prove that it is possible to distribute $p-2$ numbers to $p-2$ boys around the circle, such that the conditions of the problem are satisfied.

Let us distribute the numbers

$$2, 3, \ldots, p-2, p-1$$

to the boys around the circle in that order (see Figure 3.2).

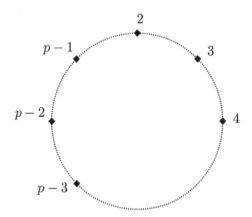

Figure 3.2   Numbers $2, 3, \ldots, p-2$ and $p-1$ are distributed among the $p-2$ in Problem 5.

Then, the second numbers are

$$2 \cdot 3, 3 \cdot 4, \ldots, (p-2)(p-1), (p-1) \cdot 2$$

We will proceed by doing the following casework.

- If one of the first $p-3$ boys have same numbers, then

$$i(i+1) \equiv i \pmod{p}$$

  Therefore, p divides the number

$$i(i+1) - i = i^2$$

  and consequently, p divides i, which leads to a contradiction.

- If the last boy has the same numbers, then

$$(p-1) \cdot 2 \equiv p-1 \pmod{p}$$

  Therefore, p divides the number

$$2(p-1) - (p-1) = p-1$$

  which leads to a contradiction.

We conclude that all the $p-2$ boys have different numbers on their papers, and the maximum value of N is $p-2$, as desired.

Solution for Problem 6

Let $f(x)$ be the left-hand side of the given equation. Notice that the term with the largest power is x^{n-1}, whose coefficient is equal to $a_1 + \ldots + a_n$. Since $a_1 + \ldots + a_n \neq 0$, then, $f(x)$ is a polynomial of degree $n-1$.

We will proceed by doing the following casework.

- Let us assume that n is even. Then $n-1$ is odd, and by the Intermediate Value Theorem, any polynomial of odd degree has a real root. Consequently, $f(x)$ has a real root, as desired.

- Let us assume that n is odd.

  - If some number a_k is equal to zero, then $x = a_k$ is a root of $f(x)$.
  - If some numbers a_i and a_j are equal, i.e. $a_i = a_j$, where $i \neq j$, then $x = a_i$ is a root of $f(x)$.

– If all a_i are distinct and nonzero numbers, then without loss of generality let us assume that they are sorted in ascending order:

$$a_1 < a_2 < \ldots < a_n$$

Notice that

$$f(a_k) = a_k(a_k - a_2) \ldots (a_k - a_{k-1})(a_k - a_{k+1}) \ldots (a_k - a_n)$$

has the same sign as the number

$$a_k \cdot (-1)^{n-k} = (-1)^{k-1} a_k$$

However, for $n \geq 3$ among the numbers $a_1 < a_2 < \ldots < a_n$ there is at least one pair of neighbors with the same sign. Then, $f(x)$ has the values of opposite signs at these points, and therefore, there is a root of the polynomial $f(x)$ between them.

We conclude that the given equation has at least one real root, as desired.

CHAPTER 4

PRACTICE EXAM #4

Problems for Day 1

Problem 1

Given positive integer numbers m and n, such that $m^2 + n^2$ is divisible by 11. Show that $m^2 + n^2$ is divisible by 121.

Problem 2

Movie tickets have six-digit numbers (from 000000 to 999999). A ticket is called *lucky* if the sum of its first three digits is equal to the sum of its last three digits. A ticket is called *average* if the sum of all of its digits is 27. Are there more *lucky* or more *average* tickets?

Problem 3

Points D, E and F are chosen on the sides BC, CA and AB of the triangle ABC, such that AD, BE and CF are concurrent. The lines ED and EF intersect the line parallel to AC and passing through B at the points P and Q, respectively. Prove that $BP = BQ$.

Practice Exams For Junior Math Olympiads - Book 1
by Roman Kvasov, Ph.D.

35

Problems for Day 2

Problem 4

For any real numbers $x, y > 1$, prove that

$$\frac{x^2}{y-1} + \frac{y^2}{x-1} \geq 8$$

Problem 5

Is it possible to cover a table $m \times n$ with some number of rectangles 1×4 and 2025 L-shaped figures consisting of 4 squares?

Problem 6

Given a circle ω_1, point A inside ω_1, and point B outside ω_1. Consider all possible triangles BXY, such that X and Y lie on ω_1 and A lies on the chord XY. Show that the circumcenters of the triangles BXY lie on the same line.

Solutions for Day 1

Solution for Problem 1

Let us consider the following table of residues modulo 11:

$x \pmod{11}$	$x^2 \pmod{11}$
0	$0^2 \equiv 0$
1	$1^2 \equiv 1$
2	$2^2 \equiv 4$
3	$3^2 \equiv 9$
4	$4^2 \equiv 5$
5	$5^2 \equiv 3$
6	$6^2 \equiv 3$
7	$7^2 \equiv 5$
8	$8^2 \equiv 9$
9	$9^2 \equiv 4$
10	$10^2 \equiv 1$

From this table we see that a square of an integer that is not divisible by 11 gives remainders 1, 3, 4, 5 or 9 modulo 11. Notice also that no combination of two nonzero remainders adds up to 11.

This implies that if at least one of the numbers a and b is not divisible by 11, then $a^2 + b^2$ will not give a remainder 0 modulo 11. Therefore, both numbers a and b are divisible by 11.

From here

$$a = 11a_1$$
$$b = 11b_1$$

for some integers a_1 and b_1.

Therefore

$$a^2 + b^2 = (11a_1)^2 + (11b_1)^2 = 121a_1^2 + 121a_1^2 = 121\left(a_1^2 + b_1^2\right)$$

and consequently, $a^2 + b^2$ is divisible by 121, as desired.

Solution for Problem 2

Answer: the number of *lucky* tickets is equal to the number of *average* tickets.

The main idea of the solution is to work with the "complements" of three-digit numbers, such that the sum of their digits add up to the number 27.

Let A be the set of all possible strings of six digits. Let X be the set of all *lucky* strings and Y be the set of all *average* strings. Notice that X and Y are the subsets of A.

Let $[abcdef]$ be some string in X, where a, b, c, d, e, f are digits. Therefore

$$a + b + c = d + e + f$$

Notice that then

$$(9 - a) + (9 - b) + (9 - c) + d + e + f = 27 - (a + b + c) + (d + e + f) = 27$$

and the string $[(9 - a)(9 - b)(9 - c)def]$ belongs to Y.

Let us now consider the mapping $f : X \to Y$, such that for all strings $[abcdef]$ from X :

$$f([abcdef]) = [(9 - a)(9 - b)(9 - c)def]$$

The mapping f is a bijection. Therefore, the number of elements in the sets X and Y is the same.

We conclude that the number of *lucky* tickets is equal to the number of *average* tickets, as desired.

Solution for Problem 3

The solution presented below refers to Figure 4.1.

Let us start by applying the Classical Form of Ceva's Theorem to the triangle ABC and the points D, E and F:

$$\frac{AF}{FB} \cdot \frac{BD}{DC} \cdot \frac{CE}{EA} = 1 \tag{4.1}$$

Notice that the triangles AFE and BFQ are similar by the Angle-Angle Postulate. Consequently

$$\frac{AF}{BF} = \frac{AE}{BQ} \tag{4.2}$$

Also the triangles BDP and CDE are similar by the Angle-Angle Postulate. Consequently

$$\frac{BD}{CD} = \frac{BP}{CE} \tag{4.3}$$

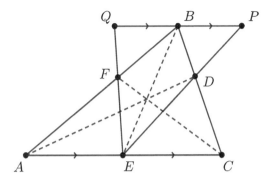

Figure 4.1 Application of the Classical Form of Ceva's Theorem to the triangle ABC and the points D, E and F in Problem 3.

Multiplying the equalities (4.2) and (4.3) we have

$$\frac{AF}{BF} \cdot \frac{BD}{CD} = \frac{AE}{BQ} \cdot \frac{BP}{CE}$$

or equivalently

$$\frac{AF}{BF} \cdot \frac{BD}{CD} \cdot \frac{CE}{EA} = \frac{BP}{BQ}$$

However, according to the equality (4.1), the left-hand side of the last equation is equal to 1. Therefore

$$1 = \frac{BP}{BQ}$$

and consequently, $BP = BQ$, as desired.

Solutions for Day 2

Solution for Problem 4

Let us make the following substitutions:

$$x = a + 1$$
$$y = b + 1$$

Then, the left-hand side of the needed inequality becomes

$$\frac{x^2}{y-1} + \frac{y^2}{x-1} = \frac{(a+1)^2}{b} + \frac{(b+1)^2}{a}$$

$$= \frac{a^2 + a + a + 1}{b} + \frac{b^2 + b + b + 1}{a}$$

$$= \frac{a^2}{b} + \frac{a}{b} + \frac{a}{b} + \frac{1}{b} + \frac{b^2}{a} + \frac{b}{a} + \frac{b}{a} + \frac{1}{a}$$

Let us now apply the AM-GM Inequality for 8 numbers:

$$\frac{a^2}{b} + \frac{a}{b} + \frac{a}{b} + \frac{1}{b} + \frac{b^2}{a} + \frac{b}{a} + \frac{b}{a} + \frac{1}{a} \geq 8\sqrt[8]{\frac{a^4 b^4}{a^4 b^4}} = 8$$

as desired.

Solution for Problem 5

Answer: no.

Let us assume that it is possible to cover a table $m \times n$ with some number of rectangles 1×4 and 2025 L-shaped figures consisting of 4 squares (see Figure 4.2).

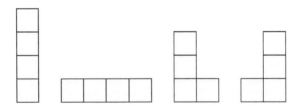

Figure 4.2 Rectangles 1×4 and L-shaped figures consisting of 4 squares used in Problem 5.

Since each of the figures contains 4 squares, then the total number of squares should be divisible by 4. This implies that either the number of rows or the number of columns is even.

Without loss of generality let us assume that the number of rows is even. Let us color the rows in black and white alternatively (see Figure 4.3).

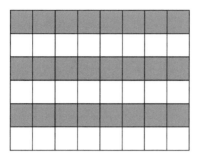

Figure 4.3 Rows of a table $m \times n$ are colored in black and white alternatively in Problem 5.

Notice that the number of black squares is even. Since each 1×4 rectangle covers an even number of black squares and each L-shaped figure covers an odd number of black squares, then the number of L-shaped figures should be even.

We obtained a contradiction. Consequently, we conclude that no such covering is possible, as desired.

Solution for Problem 6

The solution presented below refers to Figure 4.4.

Let ω be the circumcircle of the triangle BXY and let O be its circumcenter. Let Z be the projection of the point O onto the line AB. We will prove that the circumcenters of the triangles BXY lie on the line perpendicular to AB and passing through the point Z.

By the power of the point A with respect to the circle ω_1 we have

$$AX \cdot AY = OA^2 - OB^2$$
$$AX \cdot AY = (OZ^2 + ZA^2) - (OZ^2 + ZB^2)$$
$$AX \cdot AY = \cancel{OZ^2} + ZA^2 - \cancel{OZ^2} - ZB^2$$
$$AX \cdot AY = ZA^2 - ZB^2$$
$$AX \cdot AY = (ZA - ZB) \cdot (ZA + ZB)$$
$$AX \cdot AY = (ZA - ZB) \cdot AB$$
$$AX \cdot AY = (2ZA - AB) \cdot AB$$

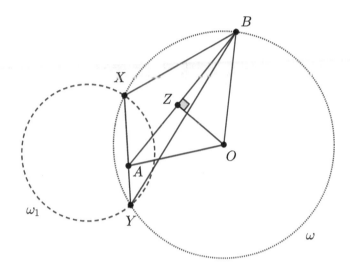

Figure 4.4   Power of the point A with respect to the circles ω_1 and ω is used in Problem 6.

From here

$$\frac{AX \cdot AY}{AB} = 2ZA - AB$$

$$\frac{AX \cdot AY}{AB} + AB = 2ZA$$

$$\frac{AX \cdot AY + AB^2}{2AB} = ZA$$

Notice that AB is constant. Also $AX \cdot AY$ equals to the power of the point A with respect to the circle ω, and therefore, is constant. This implies that the length of the segment ZA is constant, and then the point Z is a fixed point on the segment AB. From here the circumcenters of the triangles BXY lie on the line perpendicular to AB and passing through the point Z, as desired.

CHAPTER 5

PRACTICE EXAM #5

Problems for Day 1

Problem 1

Distinct real numbers a, b and c satisfy the equalities

$$a^2 + b = b^2 + c = c^2 + a$$

Find all possible values of the expression $(a + b)(b + c)(c + a)$.

Problem 2

Inside a square of area 42, there are several circles, the sum of whose circumferences is 2024. Prove that there is a line that intersects at least one hundred of these circles.

Problem 3

Find all positive integers n, for which there exists a prime p and a positive integer m, such that

$$m^n = 3^p + 4^p$$

Practice Exams For Junior Math Olympiads - Book 1
by Roman Kvasov, Ph.D.

43

Problems for Day 2

Problem 4

Let O be the point of intersection of the diagonals of the quadrilateral $ABCD$ and $AO = OC$. It is known that $\angle BKC = 2\angle ABD$. Point K is chosen on the extension of the diagonal BD beyond point D, such that $AD = CK + KD$. Prove that $\angle ADB = 2\angle DBC$.

Problem 5

Find all positive integers $m, n \leq 1997$, such that

$$m^4 + 1996 = 1998n^4$$

Problem 6

Find all functions $f : \mathbb{Z} \to \mathbb{Z}$, satisfying the following property: if a, b, c are integers, such that $a + b + c = 0$, then

$$f(a) + f(b) + f(c) = a^2 + b^2 + c^2$$

Solutions for Day 1

Solution for Problem 1

Answer: $(a + b)(b + c)(c + a) = -1$.

Let us rewrite the first equality as follows:

$$a^2 + b = b^2 + c$$
$$a^2 - b^2 = c - b$$
$$(a - b)(a + b) = -(b - c) \tag{5.1}$$

Similarly, we obtain
$$(b - c)(b + c) = -(c - a) \tag{5.2}$$
and
$$(c - a)(c + a) = -(a - b) \tag{5.3}$$

Multiplying the equalities $(5.1) - (5.3)$ we have

$$(a - b)(b - c)(c - a)(a + b)(b + c)(c + a) = -(a - b)(b - c)(c - a)$$

Since a, b and c are distinct, then we can cancel the terms $(a - b)$, $(b - c)$ and $(c - a)$:

$$(a - b)(b - c)(c - a)(a + b)(b + c)(c + a) = -(a - b)(b - c)(c - a)$$
$$\cancel{(a - b)}\cancel{(b - c)}\cancel{(c - a)}(a + b)(b + c)(c + a) = -\cancel{(a - b)}\cancel{(b - c)}\cancel{(c - a)}$$
$$(a + b)(b + c)(c + a) = -1$$

as desired.

Solution for Problem 2

Let $ABCD$ be the given square, and let $d_1, d_2, \ldots, d_n$ be the diameters of the circles. Then we have the following equality

$$\pi d_1 + \pi d_2 + \ldots + \pi d_n = 2024$$
$$\pi (d_1 + d_2 + \ldots + d_n) = 2024$$
$$d_1 + d_2 + \ldots + d_n = \frac{2024}{\pi} \tag{5.4}$$

Let us project the circles onto the side AB of the square $ABCD$. The equality (5.4) implies that the sum of lengths of the projections is equal to $2024/\pi$.

Notice that the side of the square is equal to $\sqrt{42}$. It is not hard to check that

$$\frac{2024}{\sqrt{42}\pi} > 99$$

Therefore, by the Pigeonhole Principle there exists a point X on the side AB that was projected into at least 100 times. Let us draw the line l through the point X and perpendicular to the side AB (see Figure 5.1).

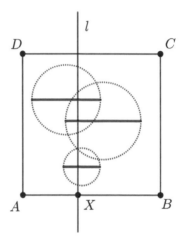

Figure 5.1 Circles are projected onto the side AB of the square $ABCD$ in Problem 2.

Consequently, the line l will intersect at least one hundred of the given circles, as desired.

Solution for Problem 3

Answer: $n = 1$ or $n = 2$.

Start by noticing that $n = 1$ satisfies the conditions of the problem.

From now on we will assume $n \geq 2$ and there exists a prime p and a positive integer m, such that

$$m^n = 3^p + 4^p \tag{5.5}$$

We will proceed by doing the following casework.

- Let us assume that $p = 2$. Then the equality (5.5) becomes

$$m^n = 3^2 + 4^2$$
$$m^n = 5^2$$

Consequently, we obtain the solution $n = 2$ that is reached for $m = 5$.

- Let us assume that $p = 7$. Then the equality (5.5) becomes

$$m^n = 3^7 + 4^7$$
$$m^n = 7^2 \cdot 379$$

where 379 is prime. Since the right-hand side is not a power of an integer, then there are no solutions in this case.

- Let us assume that p is an odd prime distinct from 7. Let us factor the right-hand side of the equality (5.5) as follows:

$$m^n = (3 + 4)\left(3^{p-1} - 3^{p-2} \cdot 4 + \ldots - 3 \cdot 4^{p-2} + 4^{p-1}\right) \qquad (5.6)$$

Let us put

$$k = 3^{p-1} - 3^{p-2} \cdot 4 + \ldots - 3 \cdot 4^{p-2} + 4^{p-1}$$

Then, the equality (5.6) becomes

$$m^n = 7k \qquad (5.7)$$

We will work modulo 7.

Notice that the right-hand side of the equality (5.7) is divisible by 7, and therefore, so should be the left-hand side. Since 7 is prime, then m is divisible by 7. Since $n \geq 2$, then m^n is divisible by at least 7^2. This implies that the right-hand side is divisible by 7^2 as well.

Consequently, k should be divisible by 7. However

$$k \equiv 3^{p-1} - 3^{p-2} \cdot 4 + \ldots - 3 \cdot 4^{p-2} + 4^{p-1}$$
$$\equiv 3^{p-1} - 3^{p-2} \cdot (-3) + \ldots - 3 \cdot (-3)^{p-2} + (-3)^{p-1}$$
$$\equiv 3^{p-1} + 3^{p-1} + \ldots + 3^{p-1} + 3^{p-1}$$
$$\equiv p3^{p-1} \pmod{7}$$

Since p is a prime distinct from 7, then we obtained a contradiction, and consequently, there are no solutions in this case.

We conclude that the only values of n that satisfy the conditions of the problem are $n = 1$ or $n = 2$.

Solutions for Day 2

Solution for Problem 4

The solution presented below refers to Figure 5.2.

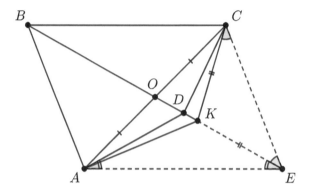

Figure 5.2   Parallelogram $ABCE$ is constructed in Problem 4.

Let us construct the parallelogram $ABCE$, by extending the diagonal BD beyond point D to point E.

Let us put $\angle ABD = \alpha$. Then $\angle BKC = 2\alpha$ and

$$\angle KEC = \angle BEC = \angle ABE = \angle ABD = \alpha$$

Since $\angle BKC$ is the external angle of the triangle CKE, then

$$\angle KCE = \angle BKC - \angle KEC = 2\alpha - \alpha = \alpha$$

This implies that the triangle CKE is isosceles and $CK = EK$. Therefore

$$AD = CK + KD = EK + KD = ED$$

This implies that the triangle ADE is isosceles and $\angle DAE = \angle DEA$. Since $\angle ADB$ is an external angle of the triangle ADE, then

$$\angle ADB = \angle DAE + \angle DEA = 2\angle DEA = 2\angle DBC$$

as desired.

Solution for Problem 5

Answer: no solution.

Let us assume that there exist integer numbers $m, n \leq 1997$, such that

$$m^4 + 1996 = 1998n^4 \tag{5.8}$$

We will work modulo 5.

Let x be an integer. Notice that if $x \equiv 0 \pmod 5$, then

$$x^4 \equiv 0 \pmod 5$$

Also, if $x \not\equiv 0 \pmod 5$, then by Fermat's Little Theorem

$$x^4 \equiv 1 \pmod 5$$

This implies that the fourth power of an integer is always congruent to either 0 or 1 modulo 5.

Therefore, the left-hand side of the equality (5.8) is congruent to

$$m^4 + 1996 \equiv m^4 + 1 \equiv 1 \text{ or } 2 \pmod 5$$

However, the right-hand side of the equality (5.8) is congruent to

$$1998n^4 \equiv 3n^4 \equiv 0 \text{ or } 3 \pmod 5$$

We obtained a contradiction, and consequently, there are no integer solutions to the given equation.

Solution for Problem 6

Answer: $f(n) = n^2 + pn$, where $p \in \mathbb{Z}$.

Let $f(n)$ be a function that satisfies the equation

$$f(a) + f(b) + f(c) = a^2 + b^2 + c^2 \tag{5.9}$$

We start by observing that the function $f(n) = n^2 + pn$, indeed satisfies the equation (5.9).

Let us start by substituting $a = b = c = 0$:

$$f(0) + f(0) + f(0) = 0^2 + 0^2 + 0^2$$
$$3f(0) = 0$$
$$f(0) = 0$$

Now let us substitute $a = x$, $b = -x$ and $c = 0$:

$$f(x) + f(-x) + f(0) = x^2 + (-x)^2 + 0^2$$
$$f(x) + f(-x) = 2x^2 \tag{5.10}$$

Let us substitute $x = a + b$ into the equality (5.10):

$$f(a + b) + f(-a - b) = 2(a + b)^2$$
$$f(-a - b) = 2(a + b)^2 - f(a + b)$$

Let us now substitute $c = -a - b$ into the equality (5.9):

$$f(a) + f(b) + f(-a - b) = a^2 + b^2 + (-a - b)^2$$
$$f(a) + f(b) + 2(a + b)^2 - f(a + b) = a^2 + b^2 + (a + b)^2$$

which can be simplified to

$$f(a + b) = f(a) + f(b) + 2ab \tag{5.11}$$

Let us put $b = 1$ and $a = n$ in the equality (5.11):

$$f(n + 1) = f(n) + f(1) + 2n$$

and let us put $p + 1$ for the value of $f(1)$:

$$f(n + 1) = f(n) + 2n + p + 1 \tag{5.12}$$

which holds for all $n \in \mathbb{Z}$.

Now we will treat positive and negative n differently. For positive n we will prove by induction that

$$f(n) = n^2 + pn$$

for all $n \in \mathbb{N}$.

The basis of induction is true for $n = 0$, since putting $n = 0$ into the equality (5.12) we have

$$f(0) = (0)^2 + p(0) = 0$$

For the inductive step let us assume that $f(k) = k^2 + pk$ for some $k \in \mathbb{N}$. Then, putting $n = k$ into the equality (5.12) we have

$$f(k + 1) = f(k) + 2k + p + 1$$
$$= k^2 + pk + 2k + p + 1$$
$$= (k + 1)^2 + p(k + 1)$$

Consequently, for all $n \in \mathbb{N}$:

$$f(n) = n^2 + pn$$

For negative n we substitute $x = n$ into the equality (5.10), where $n \in \mathbb{N}$:

$$f(n) + f(-n) = 2n^2$$
$$n^2 + pn + f(-n) = 2n^2$$
$$f(-n) = n^2 - pn$$
$$f(-n) = (-n)^2 + p(-n)$$

Consequently, for all $n \in \mathbb{Z}$ we have

$$f(n) = n^2 + pn$$

as desired.

CHAPTER 6

PRACTICE EXAM #6

Problems for Day 1

Problem 1

Distinct real numbers a, b, c and d are such that $a^2 + b^2 = c^2 + d^2$. Prove that the following equation has no positive solutions for x:

$$(x + a)^2 + (x + b)^2 = (x + c)^2 + (x + d)^2$$

Problem 2

Table 10×10 is filled with zeros. In one move, you can add one to all numbers in one row or in one column. After several moves the numbers on one of the diagonals are all equal, and are not less than any of other numbers in the table. Prove that all numbers are equal.

Problem 3

CH is the altitude in the right triangle ABC ($\angle C = 90°$). Let HA_1 and HB_1 be the angle bisectors of angles $\angle CHB$ and $\angle CHA$, respectively. Let E and F are the midpoints of the segments HB_1 and HA_1, respectively. Prove that lines AE and BF intersect on the angle bisector of angle $\angle ACB$.

Practice Exams For Junior Math Olympiads - Book 1
by Roman Kvasov, Ph.D.

53

Problems for Day 2

Problem 4

Given a positive integer number n, such that $n + 1$ is divisible by 3. Prove that the sum of all divisors of n is divisible by 3.

Problem 5

On the semicircle with diameter AC, an arbitrary point B is chosen, distinct from A and C. Let M and N be the midpoints of the chords AB and BC, respectively, and let P and Q be the midpoints of the smaller arcs formed by these chords. Lines AQ and BC intersect at the point K, and lines CP and AB intersect at the point L. Prove that the lines MQ, NP, and KL are concurrent.

Problem 6

Identical coins are distributed into several piles, each pile containing at least one coin. You are allowed to take any two piles that together contain an even number of coins and redistribute them, such that each of the two piles has an equal number of coins. A distribution is called *fair* if, through zero or more such moves, it is possible to make all piles contain an equal number of coins. Find all positive integers n for which, given any positive integer k, any distribution of kn coins among n piles is *fair*.

Solutions for Day 1

Solution for Problem 1

Let us assume that the given equation has some positive solution x. Taking into account that $a^2 + b^2 = c^2 + d^2$, let us rewrite the equation as follows:

$$(x + a)^2 + (x + b)^2 = (x + c)^2 + (x + d)^2$$
$$2x^2 + 2x(a + b) + (a^2 + b^2) = 2x^2 + 2x(c + d) + (c^2 + d^2)$$
$$2x(a + b) = 2x(c + d)$$
$$a + b = c + d$$
$$a - c = d - b \tag{6.1}$$

Let us now rewrite the equation $a^2 + b^2 = c^2 + d^2$ as

$$a^2 - c^2 = d^2 - b^2$$
$$(a - c)(a + c) = (d - b)(d + b) \tag{6.2}$$

Since a and c are distinct, the $a - c \neq 0$ and $d - b \neq 0$. Therefore, from the equalities (6.1) and (6.2), we have

$$a + c = d + b \tag{6.3}$$

Now adding the equalities (6.1) and (6.3), we obtain that $a = d$, which leads to a contradiction.

We conclude that the equation has no positive solutions for x, as desired.

Solution for Problem 2

Without loss of generality let us assume that the diagonal is going from the bottom left to the top right corners.

Let us enumerate the columns from left to right and the rows from top to bottom, and let $a(i, j)$ be the numbers in the table, where $1 \leq i, j, \leq 10$ (see Figure 6.1).

Let c_i and r_j be the number of times 1 was added to the elements of the column i and row j, respectively.

Notice that according to this notation

$$a(i, j) = r_i + c_j$$

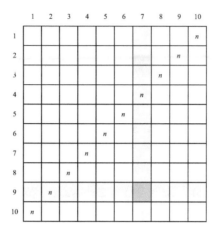

Figure 6.1   Columns are enumerated left to right and rows top to bottom in Problem 2.

It is also given that

$$c_1 + r_1 = c_2 + r_2 = \ldots = c_{10} + r_{10} = n$$

where $n \geq a(i,j)$.

Let us now assume that there exists some number $a(x,y)$ in the table that is strictly less than n, i.e.

$$r_x + c_y < n$$

We also have

$$r_y + c_x \leq n$$

By adding these inequalities we have

$$r_x + c_y + r_y + c_x < 2n$$

However,

$$r_x + c_y + r_y + c_x = (r_x + c_x) + (r_y + c_y) = 2n$$

We obtained a contradiction. Consequently, all numbers on the board are equal, as desired.

Solution for Problem 3

The solution presented below refers to Figure 6.2.

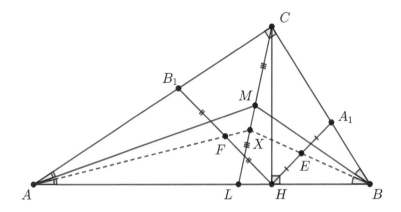

Figure 6.2   Points M and X are isogonal conjugates in Problem 3.

Let CL the angle bisector of the triangle ABC and let X be the intersection of the lines AE and BF. We will prove that X lies on CL.

Let M be the midpoint of CL. Notice that the triangles ABC, ACH, and CBH are similar by the Angle-Angle Postulate. From here $\angle BAF = \angle MAC$ and $\angle ABE = \angle MBC$.

This implies that the lines AF and AM are isogonal, and the lines BE and BM are isogonal.

From here the points M and X are isogonal conjugates. Therefore, the lines CX and CM are also isogonal.

However, since M lies on the angle bisector CL, then X also lies on CL, as desired.

Solutions for Day 2

Solution for Problem 4

Start by noticing that since $n + 1$ is divisible by 3, then

$$n \equiv 2 \pmod 3$$

Let us prove that n is not a perfect square. Let us consider the following table of residues modulo 3:

$x \pmod 3$	$x^2 \pmod 3$
0	$(0)^2 \equiv 0$
1	$(1)^2 \equiv 1$
2	$(2)^2 \equiv 1$

From here we see that a square of an integer is always congruent to 0 or 1 modulo 3. Since $n \equiv 2 \pmod 3$, then n is not a perfect square.

Let a and b be two distinct positive integers, such that

$$ab = n$$

Since $n \equiv 2 \pmod 3$, then $a \equiv 1 \pmod 3$ and $b \equiv 2 \pmod 3$, or $a \equiv 2 \pmod 3$ and $b \equiv 1 \pmod 3$. In either case

$$a + b \equiv 1 + 2 \equiv 0 \pmod 3$$

This implies that we can split all divisors of n in pairs, where the sum of the numbers in each pair is divisible by 3. Consequently, the sum of all divisors of n is divisible by 3, as desired.

Solution for Problem 5

The solution presented below refers to Figure 6.3.

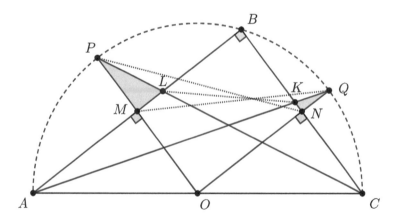

Figure 6.3 Application of the Desargues' Theorem to the triangles PML and QNK in Problem 5.

Start by noticing that the points P and Q lie on the perpendicular bisectors to the segments AB and BC, respectively. This implies that PM and QN intersect at the center O of the semicircle.

Also the lines PM and QN, ML and KQ, PL and KN, intersect at the points O, A and C, which are collinear. Now by applying the Desargues' Theorem to the triangles PML and QNK we conclude that the lines MQ, NP, and KL are concurrent, as desired.

Solution for Problem 6

Answer: $n = 2^L$, where L is any nonnegative integer.

First, notice that the operation is applied only to the piles with the number of stones of the same parity.

Let the values in the piles be x_1, x_2, ... , x_n. Let us consider the quantity S defined as
$$S = x_1^2 + x_2^2 + \ldots + x_n^2$$
Notice that under the applied operation the quantity S is a not increasing. Indeed, if the operation is applied to the piles with the number of stones a and b, then

$$\left(\frac{a+b}{2}\right)^2 + \left(\frac{a+b}{2}\right)^2 = 2\left(\frac{a+b}{2}\right)^2$$
$$= \frac{a^2 + 2ab + b^2}{2}$$
$$\leq \frac{2a^2 + 2b^2}{2}$$
$$\leq a^2 + b^2$$

and the equality holds only for the case $a = b$.

Since $S \geq 0$, then this process is finite if and only if at every step there is a pair of piles with different amounts of stones of the same parity. The process stalls if all the piles with the number of stones of the same parity have equal number of stones, i.e. if all odd x_i are equal and all even x_j are equal.

Let us show that n being any power of 2 satisfies the conditions of the problem. Let us assume that the all piles with the number of stones of the same parity have equal number of stones. Let $n = 2^L$ and let m and $2^L - m$ be the number of piles with odd number of stones $2b+1$ and even number of stones $2a$ respectively. Notice that $m \leq 2^L$. We have

$$(2b + 1) \cdot m + (2a) \cdot \left(2^L - m\right) = 2^L \cdot k$$

and consequently
$$(2b + 1 - 2a) \cdot m = 2^L \cdot (k + 2am)$$

Since $2b + 1 - 2a$ is odd, then $2^L | m$ and therefore $m = 2^L$. This means that all numbers are equal and the distribution is *fair*.

Let $n = 2^L \cdot q$, where q is an odd integer greater than 1. Let us put $k = 2$ and the total number of stones in all piles thus is $2q \cdot 2^L$. Let us consider $2^L \cdot q - 2^L$ piles with only 1 stone and 2^L piles with $q + 1$ stones:

$$1, 1, \ldots, 1, q + 1, q + 1, \ldots, q + 1$$

Therefore, we have

$$1 \cdot \left(2^L \cdot q - 2^L\right) + (q + 1) \cdot 2^L = 2q \cdot 2^L$$

Since $q + 1$ is even, it is now obvious that we cannot only perform the operations on the piles with 1 stone or on the piles with $q + 1$ stones, which does not change the configuration. Therefore, this distribution is not *fair*.

We conclude that only $n = 2^L$, where L is any nonnegative integer, satisfy the conditions of the problem.

Made in United States
Troutdale, OR
10/02/2024

23309764R00037